THE POWER OF CARING

Featuring the story of George Washington Carver

Authors
Maurine Phillips
Phyllis Colonna

Art Illustrator
Stephen P. Krause

Editor, Layout and Research
Beatrice W. Friel

THE POWER OF CARING

Featuring the story of George Washington Carver

Advisors
Paul and Millie Cheesman
Mark Ray Davis
Rodney L. Mann, Jr.
Roxanne Shallenberger
Dale T. Tingey

Publisher
Steven R. Shallenberger

Director and Correlator
Lael J. Woodbury

AN EAGLE SYSTEMS
INTERNATIONAL
PUBLICATION
ANTIOCH, CALIFORNIA

Dedicated to the children of today who will
provide the caring hearts of tomorrow.

GEORGE WASHINGTON CARVER

George Washington Carver was born in Diamond Grove, Missouri, sometime in the early 1860s to a slave mother named Mary. His father, also a slave, was killed in an accident shortly after his birth. He had one older brother, Jim.

George's mother belonged to Moses and Susan Carver, German immigrants to the American frontier. When George was a few months old, he and Mary were stolen and carried into Arkansas by slave raiders. Mary was never heard from again, but the baby was later ransomed to the Carvers for a racehorse.

The frail child grew up on the Carver plantation with a keen eye for plants and animals. He had a delicate sense of line and color and taught himself to paint. Later he became well-known among the blacks as a singer and organist.

Although the Carvers told George he was no longer a slave, he chose to stay with them until he was about ten years old. Then he left to acquire an education. For the next fifteen to twenty years he pursued knowledge through every means he could find, traveling through many different states. He attended grade schools in four different towns in Missouri and Kansas, supporting himself by whatever work came to hand.

He finished high school in Minneapolis, Kansas, when he was in his late twenties. Refused admittance to a Kansas university because he was black, for five years George gave up his dreams of a higher education. Friends finally persuaded him to apply for admittance to Simpson College, Iowa, where he studied piano and art. He then attended Iowa State Agricultural College and received a degree in agricultural science in 1894 and a master of science degree in 1896.

In 1896 Booker T. Washington called him to Tuskegee Institute to head the school's newly organized department of agriculture. Here George began a series of creative experiments that brought him international fame. Despite many offers to go elsewhere—including one from Joseph Stalin to superintend cotton plantations in southern Russia—he stayed at Tuskegee because he believed in its destiny as a learning institution for black people.

He taught Southern farmers to plant soil-enriching peanuts and sweet potatoes instead of soil-exhausting cotton. Then he sought new uses for the surplus foodstuffs. He developed three hundred synthetic products from peanuts alone, including cheese, flour, ink, dyes, plastics, soap, linoleum, and cosmetics. When he arrived at Tuskegee, the peanut had not yet been recognized as a crop. Within fifty years it was one of the six leading crops in the United States.

A deeply religious man, George Washington Carver was criticized for his humility (some called it subservience) and for his belief that God was his partner in the laboratory. In truth he was a gentle, soft-spoken, modest man whose greatest desire was simply to serve humanity—his own people, the poor blacks of the South, in particular. He gave his discoveries freely to mankind, refusing large sums for their commercial exploitation. He declined an offer to work for Thomas Edison at a salary of more than $100,000 per year.

In 1937 a bronze bust of George Washington Carver was unveiled in his honor at Tuskegee Institute. It was paid for by one-dollar subscriptions sent in by his many admirers, black and white. In 1953, ten years after his death, his birthplace was declared a national monument. The words of Mariah Watkins had proved prophetic:

That boy told me he came to Neosho to find out what made hail and snow, and whether a person could change the color of a flower by changing the seed. I told him he would never find all that out in Neosho nor in Joplin—maybe not even in Kansas City. But all the time I knew he'd find it out somewhere.

"Here now, Missy—what are you doing?" Mr. Possum asked the tearful little girl. "Why are you trying to hide behind that ridiculously small tree, and why are you making those sad sounds? Don't you know that upsets me?"

"Go away!" the little girl answered. "You're just an animal, and animals can't talk!"

"Why, I am not just an animal!" Mr. Possum answered, shocked to the tip of his nose. "There is no such thing as *just an animal*. And I *can* talk. My name is Percival Possum. Percival Augustus Possum, to be exact. I would like to help you, if you'll stop dripping tears into that soggy handkerchief and tell me why you're crying."

The little girl sniffed once or twice, then wiped her eyes. "I don't have any friends," she said at last. "Nobody cares about me at all."

"What nonsense," Mr. Possum replied. "Of course someone cares. I care. Turn around here now and let's have a little chat. Are you certain you don't have any friends? Maybe you don't care enough about other people. I have found that a caring heart—a truly caring heart—can make wonderful things happen."

"But we just moved here," the little girl said, her tears starting again. "The boys and girls in school don't want to play with me. They say I talk funny and I'm different."

SNIF !!

"Oh dear, dear," Mr. Possum murmured, quite distressed. "People aren't always as thoughtful as they might be. Usually it's because they don't understand, not because they don't like someone. Anyway, they're right you know. You *are* different. Every person is different from every other person. That's what makes the world such an exciting place to live."

The little girl sighed, "Well, it doesn't seem very exciting to me," she said. "It just seems like a whole lot of trouble."

Mr. Possum studied her thoughtfully over the tops of his spectacles. "I could tell you about someone who knew real trouble," he said. "He was different. He was sickly and poor, and he was born a slave. Yet, he had a caring heart, and he became very important to all the people in the South. In fact, he became important to the whole world. Would you like to hear his story?"

The little girl nodded her head, and Mr. Possum settled himself comfortably on the grassy bank. If there was anything he enjoyed, it was the telling of a good story.

"Well now, Missy, we'll have to travel backward through time to about 1860," he began. "Sometime in the early 1860s a baby boy was born in Diamond Grove, Missouri. The log cabin floor was made of dried mud, and the cracks in the log walls were filled with mud and straw to keep out the winter wind."

"It was no usual pioneer family that lived in this home. Mary, the baby's 19-year-old mother, was a slave. The baby's father was dead. He had been a slave, too, on the neighboring farm. Mary and her two little sons belonged to a man named Moses Carver. This baby boy would be known as Carver's George."

"Now you mustn't think Moses Carver was an unkind man by nature. Neither he nor his wife, Susan, believed in slavery, but it was a fact of life in the times and places where they lived. They were immigrants to Missouri, and building a home in the wilderness was hard work. They were beginning to grow old, and Moses thought Susan needed a companion and helper. They paid $700 for Mary when she was only thirteen.

"I know it isn't right, Susan," Moses said. "But at least if she's with us, we can see that she has a good home. That's more than she might have with someone else."

"Mary did have a good home with the Carvers; they came to love her very much. But those were bad days in Missouri. Feelings over the question of slavery ran strong. Half the people believed in it, and the other half didn't. Many men used violence to force their ideas on others. Mobbings and burnings were common, forcing some farmers to leave their farms to move to other places. Many other farmers kept their families and farm animals hidden in the woods for safety. People couldn't take care of their fields and gardens as they needed to, and it seemed that most folks were almost always hungry."

"George wasn't strong when he was born, and sickness and hunger made him even weaker. One cold night Mary sat close to the fire cradling her baby in her arms. The flame was so small that it scarcely held back the chill of the room. She sang softly to quiet George's crying, and now and then she spooned a few drops of honey into his mouth. It seemed to soothe his cough."

"Suddenly the door burst open. *"Slave raiders!"* Moses Carver cried. *"Run, Mary!"* He scooped the oldest little boy into his arms and raced for the thick woods. But Mary wasn't fast enough. She and little George were thrown onto a horse, and Moses and Susan watched them vanish into the night."

17

"The kind old couple spent many anxious days wondering if they would ever see Mary and George again. Moses offered some of his best timberland and a good horse to anyone who could bring them back again. Some rumors said Mary had been taken north by Union soldiers. Others said that she had been sold to a plantation in Louisiana. One rumor said she was dead. Moses followed every lead, but none took him to Mary and George."

"Then one day a man named Mr. Bently rode into the yard. He unstrapped a bundle from his saddle and dropped it into Susan's arms. *"There,"* he said roughly. *"I guess it's still alive."* Carver's George had come home, but no trace was ever found of Mary."

"It took weeks of careful nursing before Susan was sure the little boy would live. Even then his illness stretched into long months. In fact, he was almost three years old before he could walk. He never did get so he could run and climb trees and work as hard as his brother, Jim, but his eyes noticed everything that went on around him. He spent hours watching the busy insects, and he often carried a worm or a beetle in his pocket."

"Oh, Ugh!" Missy shuddered. "Why would he want to do that? Bugs and worms are creepy!"

"Well, George didn't think so," Mr. Possum answered. "He wanted to understand why they do the things they do. He saw all the plants and animals and little crawling creatures that live in the woods, and he wanted to know how they help each other."

"There wasn't any piece of life he didn't care about! Questions buzzed around his head like bees around a honey tree.

"Where does rain come from?"

"What makes the snow?"

"How does a caterpillar turn into a butterfly?"

"Why is one plant healthy and strong while the plant next to it is yellow and sick?"

"Sometimes George carried bits of nature home with him—seedpods, rocks, a grasshopper, a piece of moss. But what looked like treasures to him looked like pieces of trash to Aunt Susan.

"*George, you take that right back outdoors where it belongs!*" she told him. She had a caring heart, and she was glad to see George interested in so many things. But she liked to keep a clean house, too!"

23

"So George found his own secret place deep in the woods, where warm sunlight filtered down through the tall trees. He even made a little greenhouse from moss and bark to protect his plants when winter came."

24

"You mean the Carvers let George play in the woods all by himself?" Missy asked, very surprised.

"Oh yes!" Mr. Possum chuckled. "George knew the woods as well as he knew his own cabin. He wasn't strong enough to help with the heavy work, so his job was gathering flax and hemp* for Aunt Susan to weave into cloth. Another job was looking for roots and herbs to make medicines. While he was looking and gathering, he had plenty of time to study everything around him, even the ground he walked on. He loved to feel the soil in his hands. Some folks said he could read its secrets while he let it slip through his fingers."

*Flax and hemp are plants whose fibers were used to make cloth and rope.

"The people in Diamond Grove started calling him *the little plant doctor.* They often asked him to come and take care of their sick plants. When Aunt Susan's petunias wouldn't blossom, George went out to take a look. *"All they need is a little sand added to the soil,"* he said, and soon the flowerbed was filled with bright blossoms."

"Uncle Moses was so impressed that he asked George to look at his favorite apple tree. *"It doesn't seem to be doing too well,"* Uncle Moses said. George discovered that some of its branches were filled with insects. He cut those branches off and burned them, and soon the tree was as good as new. *"I do declare!"* Uncle Moses went around saying. *"There isn't anything that boy doesn't know!"*

THINK ABOUT IT

1. What were some of the ways Moses and Susan showed that they really cared about George and his mother?
2. What are some of the ways George showed that he cared about Susan and Moses?

"George knew there were many things he didn't know, and he wanted to learn every one of them. One day he had an errand to run that took him past a little log building near Diamond Grove. He heard voices inside, and when he looked through the window, he saw a man talking to a group of children. George had never seen or heard of a school before. The more he listened, the more excited he got. He ran all the way home."

"Aunt Susan! Uncle Moses!" he called. "There's a place called a school, where boys and girls can learn to read and write. I've been watching them! I want to learn, too. May I go to school tomorrow?"

"Moses and Susan looked at each other. They felt very sad. Uncle Moses put his arm across George's thin shoulders. "I'm sorry, George," he said. "Black children aren't allowed to go to school. You'd best just forget about it."

29

"But George couldn't forget about it. He felt so bad that Aunt Susan decided to give him school lessons at home. She took a piece of burned wood from the fireplace and wrote the letters of the alphabet on the fireplace rocks. Uncle Moses wanted to help, too. He taught George how to write numbers and how to add and subtract. Soon George could read, write, and do arithmetic as well as Moses and Susan."

"But that still wasn't enough for George. It wasn't just learning to work with his mind that he cared about. He wanted to learn to work with his hands, too. Everytime he saw someone doing something that looked hard, he would say to himself, *"Well, they do that with their own two hands. I've got hands, and I'll bet I can learn to do it, too."*

"Soon he had learned to repair shoes and make furniture by watching Uncle Moses. From Aunt Susan he learned to knit, spin, cook, and make candles. It seemed he could learn anything he set his mind to."

"One day he had work that took him to a big mansion. While he was waiting in the hallway, he looked at the pictures on the walls. He hadn't seen oil paintings before, and he thought he had never seen anything so beautiful. He stared at them for a long time; then he walked over and gently traced the lines with his fingers.

"*Well, somebody made these pictures with his own two hands,*" he said at last. "*And if he can do it, then I can learn to do it, too.*"

Missy laughed. "I'll bet he did, Mr. Possum! I'll bet he learned to paint beautiful pictures, just like he said he would."

"Indeed he did, Missy," said Mr. Possum, looking very pleased. "But that isn't all. You see, George didn't have any money, so he couldn't buy paints or canvas or even brushes."

"That meant he had to find ways to make his own paints out of crushed berries or leaves and flowers. Sometimes he mixed them with oil; sometimes he mixed them with clay. Since he didn't have a brush, he learned to paint with his fingers. And since he didn't have canvas, he painted on anything he could find. Soon people were surprised to find pretty pictures painted on rocks, tree trunks, and old pieces of boards."

"But the things George learned to do at Diamond Grove still didn't seem to be enough. As the years passed, he began to feel restless. The Civil War had been over for many years, and all the slaves were free. George was free, too, but he had been staying with Uncle Moses and Aunt Susan anyway. When he was about ten years old, he decided the time had come for him to make his own way in the world.

"What's the matter, George?" Aunt Susan asked after breakfast one morning. "You seem so quiet these days."

"Aunt Susan, I hear there's a school for black people in Neosho, about a day's walk from here," George answered. "I feel like I have to go see what I can learn there."

35

"Aunt Susan didn't know what to say. She knew George had to live his own life, but she loved him and knew she would worry about him when he was gone. "*Well, we always knew you couldn't stay with us forever,*" she said at last. "*There are things for you to do that you have to find out for yourself, George. But we'll surely miss you.*"

"While George went out to tell Uncle Moses what he had decided, Aunt Susan ironed him a clean shirt and wrapped up a chunk of ham and a piece of cornbread. He could eat it along the way when he got hungry. She and Moses stood in the doorway and watched George as far as they could see him. They wondered what he would find to eat and how he would earn a living. They wondered if anyone would give him a place to stay. They knew he was a gentle boy with a caring heart. They hoped he would find other people who would care about him, too."

"As for George, he didn't think about any of those things. He just thought about how far away Neosho was. He walked until both his legs ached, and then he walked some more. It was almost dark when he arrived. When he finally found the little schoolhouse, it was locked and empty. George was so tired that he plunked himself right down on the schoolhouse steps. He was hungry, too—and more lonely than he had ever been before.

"Well, I don't want to spend the night on these hard steps," he thought as he shivered in the night air. *"But where can I spend it?"*

"Just then he heard a horse stomp and whinny. *"That came from down the street,"* he said. *"Where there are horses, there is hay. And hay is a whole lot more comfortable than where I'm sitting!"*

"George followed the sounds of the horses, and soon he had found a warm stable. He climbed up to the loft and burrowed into the clean, sweet-smelling hay."

"It seemed he had scarcely closed his eyes when the morning sun came streaming through the narrow windows. George sat up and rubbed his eyes. He was even more hungry than the night before."

"He walked back to the little schoolhouse, but it was still locked. He looked around for something to eat, but all he could find was one big sunflower. He sat down on a pile of wood and began to eat the seeds."

"Suddenly a voice yelled right behind his ear, *"Here now! What you doing on my woodpile?"*

"Now whenever George was surprised, he stuttered; and he was so surprised now that it took a few minutes before any words would come out of his mouth. A little old black woman was standing right next to him, and she looked very grouchy.

"I'm just sitting here waiting for school to start," George finally managed to say.

"The little old woman looked grouchier than ever. *"School don't start 'til Monday,"* she said. *"This here is Saturday, You thinking of perching on my woodpile for three days?"*

"George just stared at her. He hadn't known it was Saturday. He hadn't even known schools let out for the weekend.

"*You hungry?*" the little old woman asked. George could tell she wasn't really as angry as she had pretended.

"*Yes'm,*" he answered.

"*Then get yourself clean at that pump over yonder and come in and eat,*" she told him."

43

"The little black woman's name was Mariah Watkins. Even though she tried to act grumpy and grouchy, the truth is, Mariah was as kind and loving as Susan Carver. She cared about people very much, and her heart reached out to George the first minute she saw him sitting there, all hungry and forlorn.

"She introduced him to her husband, Andrew, and then fixed a breakfast of hot biscuits with butter and molasses. *"You have any place to stay?"* she asked him when he finished eating. George shook his head no."

"Well, if you're not afraid to help with the work, you can stay here," she said. "You'll be right handy to the school." George's eyes filled with tears. "I sure was lucky to pick your woodpile to sit on," he said."

THINK ABOUT IT

1. How do you think Mariah and Andrew could tell that George had been taught to care about other people?
2. How do you think they might have treated him differently if they had thought he was not a kind person? Why?

GEORGE SHARES HIS GIFTS WITH OTHERS

"George lived with Mariah and Andrew for two years and grew to love them very much. Mariah was a very religious woman. She taught George to go to church and to pray.

"I never knew anybody like you, Aunt Mariah," he said. "You talk to God like he was a good friend standing right beside you."

"Of course I do!" Mariah answered. "That's just what he is!" Soon that's the way George began to feel, too. He learned to get up every morning before dawn and walk through the hills and countryside. "The world is God's garden," he said. "That's where I like to go to talk to him." Mariah gave George an old worn Bible that he kept with him for the rest of his life. He learned whole sections of it by heart."

"The time finally came when George decided he had learned all that the little school in Neosho could teach him. One day a neighbor offered him a ride to Kansas if he would help drive the wagon and mules. George said he would go, but saying goodbye to Mariah and Andrew Watkins wasn't any easier than saying goodbye to Susan and Moses Carver had been. They were kind and loving people, and George cared for them very much."

"Aunt Mariah said a prayer as she watched the neighbor's old wagon creak down the road with George perched high on top: *"Dear Lord, get him a good school and a right smart teacher, 'cause there's an awful lot that boy wants to know."*

"For more than ten years George wandered to wherever he could find work and a new school to go to. Sometimes he worked for several months to get enough money to go to school for a few weeks. When he had learned what he could in one place, he moved on to a place where he could learn something else. He learned from books, experience, and other people. And he learned by working with his hands."

"His travels took him through Kansas, Colorado, Iowa, and New Mexico. He worked as a cook, a houseboy, a farmhand, and a laundry man. Everywhere he went he still got up before dawn to walk through God's garden and talk to his friend, just as Mariah had taught him to do. He still studied all the plants and animals of the countryside, too. He learned the names of most of them, and he began to paint them once again—only now he could afford to buy real paints and brushes."

"In Minneapolis, Kansas, George worked as a farmhand long enough to graduate from high school. He was twenty-five years old and had grown to be six feet tall. He was a lot older and taller than most of his classmates, but he was so happy to have his high school diploma that he didn't even think about that."

51

"*Now I'm going on to college!*" he told his friends. He sent an application to several universities, and finally one wrote back that he was accepted. George took all his savings and bought a ticket to where he thought he would be going to school. But when he got there, he was stopped at the door. He hadn't thought to write that he was black, and black people weren't allowed to go to that school."

"Mr. Possum, I don't understand that at all!" Missy said.

"Well now, Missy, that's something I don't understand too well myself," Mr. Possum admitted.

"As a matter of fact, George felt so bad that for five years he forgot all about his dream of going to college. But one of the nice things about having a caring heart is that other people care about you, too. George had many kind friends, and they talked him into writing to Simpson College in Iowa. The people at Simpson wrote right back, and he was accepted as a full-time student at last."

"All the teachers at Simpson were excited about George's bright mind and his many talents. His grades were high in every subject, and he took part in stage plays and other entertainment as well. He was a serious artist, and he learned to be a very good musician. But his special interests were still *all the little plants and living creatures in God's garden.*"

"George's teachers talked him into transferring to the Iowa State Agricultural College at Ames, Iowa. *"We think you have a brillant career ahead of you in the field of agriculture."* they told him.

"They were right. George graduated from Iowa State College in 1894. Then he went on to receive his master's degree in 1896. He was offered a position as a teacher at the college, but someone else had plans to use George's talents, too."

"At Tuskegee, Alabama, a black man named Booker T. Washington was struggling to start a school especially for black students. He wanted George to come and take charge of the department of agriculture. *"I can't offer you the money you would earn at Ames,"* he told George. *"I can offer you hard work and the opportunity to help our people. I can teach them how to read and write, but I can't teach them how to plant and harvest. Will you come?"*

"George remembered what Aunt Mariah had told him many years before. "God has something important for you to do, George. You go out and learn all you can, and then you find a way to teach it to our people."

"I will come," George told Booker T. Washington."

"He truly did have a caring heart, didn't he, Mr. Possum?" Missy asked.

"Oh indeed he did, Missy! And because he had a caring heart, he made wonderful things happen! He spent the rest of his life looking for ways to help the poor people in the South. He became a great teacher and a great scientist. He taught other people all the things he had learned about the soil and the way plants and animals help each other."

"You shouldn't plant cotton all the time," he told the farmers. "Cotton weakens the soil. Plant peanuts and sweet potatoes. They make the soil stronger."

"But Mr. Carver, no one wants to buy peanuts and sweet potatoes," they told him. So he went into his laboratory and found 300 new ways to use peanuts and 118 new ways to use sweet potatoes! Soon people were making as much money growing peanuts as cotton, and their farms were getting better and better."

60

"George made so many important discoveries that people came from all over the world to visit him. Henry Ford, Thomas Edison, Mahatma Gandhi, two presidents of the United States, and many foreign governments asked for his advice. Some men offered him a lot of money for his discoveries—more than $100,000 a year—but George always told them no. *"I only want to help the people,"* he said. *"Anything I discover belongs to them."*

"So you see, Missy, one man who truly cared made life better for everyone who lived in the South. And his discoveries are still helping people today in almost every country."

Missy stood up and brushed off her pretty dress. "Where are you going?" Mr. Possum asked, surprised.

"I'm going back to school," she answered decidedly. "There are a lot of things I still need to learn! There might be something important for me to do in the world! Besides, with that many students, there must be one or two of them who would like to be my friends. Especially when they find out I care about being a good friend, too."

"Hurrumph!" said Mr. Possum, trying not to show how very pleased he was. He watched Missy walk through the schoolhouse door. Then, with a satisfied shake of his round little self, he started on his way. After all, there might be other sad little girls or boys just needing someone who enjoys the telling of a good story. There might be girls and boys who need to know that someone cares.